CW01512020

Terranexus

Connection and Meaning in Ordinary Places

Terranexus

Connection and Meaning in Ordinary Places

David K. Leff

The Little Bound Books Essay Series
Personal. Poignant. Powerful.

The author has tried to recreate events, locales and
conversations from his memories of them. In order to
maintain their anonymity in some instances he has changed
the names of individuals and places, he may have changed
some identifying characteristics and details such as physical
properties, occupations and places of residence.

Published in 2018 • Little Bound Books
Imprint of Homebound Publications
Front Cover Image © by Eco Grid
Cover and Interior Designed • Leslie M. Browning
ISBN • 978-1-947003-95-8
First Edition Trade Paperback

10 9 8 7 6 5 4 3 2 1

Homebound Publications is committed to ecological
stewardship. We greatly value the natural environment and
invest in environmental conservation. Our books are printed
on paper with chain of custody certification from the Forest
Stewardship Council, Sustainable Forestry Initiative, and the
Program for the Endorsement of Forest Certification.

ALSO BY DAVID K. LEFF

The Last Undiscovered Place

Deep Travel

Hidden in Plain Sight

The Price of Water

Depth of Field

Tinker's Damn

Finding the Last Hungry Heart

Maple Sugaring

Canoeing Maine's Legendary Allagash

———

To Eric D. Lehman, fellow deep traveler,
for his encouragement,
sound advice on craft,
and most of all, his abiding friendship

It takes a man of genius to travel in his own country, in his native village; to make any progress between his door and his gate.

—Henry David Thoreau, Journal, August 6, 1851

I

SUDDEN ONSET TERRANEXUS

In the heart of New Hampshire's White Mountains, I took my first steps up the well-worn Crawford Path. It was the beginning a solo trip across the Presidential Range with stays at the legendary high huts of the Appalachian Mountain Club. Intoxicated by spicy spruce air, eager for the steep climbs promised by the map's clustered contours, I passed a yellow Forest Service sign warning of bad weather, danger and even death above tree line. This only fueled my teenage rite-of-passage dreams, and the white Appalachian Trail blazes stirred an urge to walk from

Georgia to Maine. Here on America's oldest recreational footpath, cut in 1819 by rugged hostler Abel Crawford, I felt my journey somehow connecting with generations past. It was my first adventure in the wild, and I was ready for a test of manhood. What I was not prepared for was to discover an entirely new way of experiencing landscapes.

Although the first day's hike to Mizpah Spring Hut was only two-and-half miles at a relatively easy grade, the 1,900-foot elevation gain exhausted me. I nevertheless became reenergized on reaching my destination in the windswept col between 4,000-footers Mts. Jackson and Pierce. Maybe it was the bracing alpine air, the anticipation of climbing to the bare rocky tops the next day, or just a novice's sense of accomplishment. I reveled in conversation with fellow hikers, swapping stories over the hut crew's homemade soup and fresh-baked bread, listening to old timers telling heroic and humorous tales. I also made a brief raid on the small library of well-thumbed books, soaking up history, geology, ecology and lore.

The next day offered stiff breezes and azure sky dotted with clouds. Along the ridge, astringent light revealed a roiling sea of green slopes punctuated with ashen ledges that seemed to break like ocean whitecaps. The trail moved through small trees, krummholz thickets, and lichen-crusted rocky plains where hardy mountain plants grew in tufts and low mats. I was gripped by the powerful natural grandeur, and the path marked by crude cairns and sometimes edged with stones seemed to possess the nobility and timeworn quality of a Roman road.

While my senses feasted on the on the stark and cragged terrain, my imagination wandered in time as well as space to Indian legends, grand hotels, logging railroads, wildfires, home-crushing landslides, rare flowers, and ancient continental crashes that formed the mountains. Not limited merely by what I saw at any given moment, I encountered a richer, denser world. It was love at first sight. I felt a sudden, strong attachment, manifestation of a phenomenon I've since called "terranexus," profound connection to terrain or place.

I'd fallen in love with places before, just never so suddenly. It was usually, even at such a young age, by long association—the wooded lots I wandered in the suburbs at home and my great uncle's New Jersey chicken farm among others. But this time my infatuation was clear and precipitous.

We spend time in places like the White Mountains for recreation, physical challenge and to find respite from an ever accelerating frenetic life. This "land of many uses," as the U.S. Forest Service has called our national forests, has been set aside for scientific study, wildlife, timber harvests, hunting and fishing, clean water, and to protect geological marvels and historic sites. But while these practical reasons support conservation initiatives and appropriations, they tell a junior part of the story. After all, many other places boasting such values remain unprotected. Ultimately, the political, financial, and social will to create and maintain the White Mountain National Forest and areas like it results from deep affection by small cadres of activists and wide swaths of the public.

Federal legislation, layers of management plans, wilderness designations, and all the hard work that has gone into creation and maintenance of the forest are perhaps less the result of science, resource management needs, recreational desires, and tourist dollars than it is a matter of love, attachment of people to a place. And it's not just woodsy quiet or the austere beauty of summits that stirs such passion, but a history of human interaction from Native Americans to the pioneering Crawfords, from the cog railway to the hut system to the Mount Washington Observatory.

If we are to protect wild areas as disparate as Labrador and the Amazon from invasive species, encroaching development, climate change and other threats, it will take terranexus, a revival of connection and romance between people and places. And it will require not just a fondness and devotion to distant and wild locations, but to nearby and built environments where our attitudes toward land are formed. Only by appreciating, understanding, and improving the everyday landscapes in which we live and work will we be able to ensure the future of the White Mountains and other places where nature is dominant and human beings feel like visitors. The close-to-home areas where we first learn about the outdoors "are places of initiation," ecologist and author Robert Michael Pyle has written, "where the borders between ourselves and other creatures break down, where the earth gets under our nails and a sense of place gets under our skin. They are the secondhand lands, the hand-me-down habitats where you have to look hard to find something to love."

After that first White Mountain journey, I longed to recapture the happy endorphins of falling in love with a place, the adrenalin rush of sudden onset terranexus. But with life becoming busy as I progressed into my twenties, it was increasingly tough getting to remote, untrammeled destinations. So I began exploring close to home, finding near-at-hand places more alluring and intriguing than I had ever imagined.

Maybe it was my youthful delight in doing something contrary and unexpected, but I itched to canoe western Connecticut's once famously and hopelessly polluted Naugatuck River. Though it was nearby, it was a kind of terra incognita for paddlers. No one I questioned had attempted it. "Had hepatitis shots?" I was asked half jokingly by friends who knew the river's reputation. Then, sensing I was in earnest, they stared with a chilling mix of concern for my safety and wonder at my sanity. Naugatuck River water carried disease, I was warned, hid submerged metal objects that could slit the toughest canoe hull, and was laden with unmarked dams that remained invisible until too late. With a bit of trepidation, I pondered the warnings and the opportunity for years before finally dipping my paddle in the forbidden waters.

Until little more than a generation before my journey in the early 1990s, the Naugatuck had been the nation's center of brass manufacture for over a century. Along the river's banks, the metal was forged from copper and zinc and fabricated into clock parts, kettles, buttons, buckles, tubing, hinges, artillery shells, wire, gears and automobile parts. Over time, the swift waters became an open sewer for industrial acids, dyes, heavy metals and human waste. As early as 1899, a government report found that the river had reached the limits of permissible pollution. A 1966 survey found not a single fish alive in its 39-mile length.

Once actually on the water, I found a world of startling contrasts. The river flowing past wild-seeming banks lined with silver maple, sycamore and cottonwood where black ducks, mallards, mergansers, and geese floated, also rushed through small, gritty cities with fortress-like factories sprouting stacks, vents, exhausts and intakes. At one point I passed a maze of pipes and tanks at a chemical plant and then suddenly found myself in a constricted valley of tumultuous rapids where ospreys and turkey vultures soared above steep forested ledges. Near where drowned shopping carts and piles of bald tires collected like suicides, I spied muskrats and freshly peeled beaver chewed sticks turning slowly in eddies. Sewage treatment, control of industrial discharges, and the closure antiquated factories were resulting in rebirth of a river once shunned and given up for dead. Water quality has gotten even better in the ensuing years. Over thirty species of fish are now at home in the Naugatuck.

Despite some trash along its banks, and passage through densely developed urban and suburban areas, I found the river beautiful, not only for where it went, but for where it had been and where it was going. A place that was once a joke had become a source of inspiration and fun eliciting joyful, not derisive laughter. I was struck by terranexus in this debased and degraded place, not just because of unexpected natural beauty, but because its long and tangled encounters with humanity added layers of intrigue.

My Naugatuck voyage spawned a new interest, affection almost, for the hard used places where civilization and nature are entangled. About a mile from home via a walk wholly on pavement, such a location became my "listening point," a place where I can be wholly within myself and both contemplate the grand mysteries of the universe and mundane conundrums of daily existence.

The term "listening point" was coined in the title of naturalist Sigurd Olson's 1958 book, a paean to his special place, a bare glaciated spit of rock at the water's edge in northern Minnesota. Each time he went there it "opened great realms of thought and interest" where he saw "the immensity of space and glimpsed at times the grandeur of creation." He christened his spot "listening point" because "only when one comes to listen, only when one is aware and still, can things be seen and heard."

My listening point is neither remote nor secluded. It's a mere fifteen-minute walk from where I live, a place where sometimes hundreds of people pass daily on foot, bicycles or roller blades. But, I can be there easily, at the slightest whim whenever the day gets hectic or the spirit moves me. Until little more than half a century ago, most of my walk to this spot was a corridor for locomotives hauling freight. The site itself was a hydroelectric station. But as Olson observed, a listening point does not have to be "close to the wilderness, but some place of quiet where the universe can be contemplated with awe."

From this perch on the west bank of Connecticut's Farmington River at the lower Collinsville Dam, I often sit on one of the rusted I-beams embedded in the concrete abutment that once served as a gate structure to bypass water. It's surrounded by trees, offering a commanding view of riffles downstream and the dark, glossy impoundment above. Across the water are a moldering brick gatehouse and the long concrete façade of the old power canal, both brightly tattooed with graffiti. Half hidden in the woods, they appear like remnants of a long lost civilization.

While not isolated, the spot provides ample solitude. The falls' roar is an insulating sound blocking all outside noise. The smooth water continuously rolling over the concrete spillway and plunging to a milky froth below is alluringly hypnotic and helps focus the mind. It takes but a moment to feel remote despite people nearby.

I'm energized at this oasis of tranquility. Long out of business, the axe and machete manufacturing Collins Company recognized the river's power early in the last century when it built the dam to funnel water 650 feet downstream to a powerhouse that spun twin turbines generating a combined 700 horsepower. Now, with the water unbridled by any dynamo, I can contemplate that unfettered power and let it flow through me. At the same time I share the river's force, it has a calming influence. Any restiveness or agitation easily flows downstream.

As water tumbles over the precipice and I feel its mild thunder in my chest, I often concentrate on breathing, feel my abdomen rise and fall, the chest expand and contract. The mind gains free reign and wanders like a dog out for a walk.

At this one time site of industrial power I now enjoy a place rich in natural beauty and deep ecological function where I feel the kinetic energy of moving water while gazing at the slowly deteriorating imprint of humanity. I've spent many hours here wondering about our need for a deeper mutual relationship with the natural world. Thought fragments and fleeting ideas have come to me while almost in a listening-point-dream-state. Over the years, I've pieced together a framework that helps explain, to me at least, where that relationship with nature has been and where it might be going.

"In wildness is the preservation of the world," Henry D. Thoreau uttered in one of his lightning-bolt aphorisms. The earth's wild places have clearly shrunk since his time, making the remaining ones even more precious for the homeostatic regulation of the planet. But beyond ecological value, their worth as wellsprings of human imagination have grown exponentially as their acreage has contracted. Counter intuitively, perhaps, protection of such places requires we appreciate nearby landscapes. To turn the Concord naturalist's dictum somewhat around, I'm convinced that in built spaces where we live is the preservation of the wild. Terranexus is necessary for human survival. I've heard this mantra whispered in the wind and falling water at my listening point.

II

CONSERVATION'S FOURTH WAVE

America has experienced three waves of conservation consciousness pointing the way to terranexus. While scientists, politicians, and artists, sportsmen and travelers have contributed to each wave, writers have probably been the most influential. No wonder Bill McKibben, one of today's finest environmental authors, has observed that "an argument can be made that environmental writing is America's single most distinctive contribution to the world's literature."

The first wave established an awareness of the beauty and diversity of nature, creating a realization that the natural world was not just something for exploitation. While there are antecedents, in America this approach was pioneered by Thoreau whose masterpiece, Walden, was published in 1854. The book is full of keen nature observations and acerbic critiques of society, but most significantly makes a compelling connection between human consciousness and natural objects and phenomena, the first stirrings of terranexus. To some extent, all environmental writing is a footnote to Thoreau, but there are many clear-eyed and compelling authors in this first-wave tradition of awareness. These include Catskill naturalist John Burroughs in the later nineteenth century, and Edwin Way Teale who won the Pulitzer Prize for one of his epic four volumes about seasonal change in the twentieth. Among contemporary writers are John Hanson Mitchell who interweaves human culture with nature, and time with place. Another is Robert Finch, whose Cape Cod essays tie personal experience with history and nature's cycles.

The second wave launched from the first. It called not just for awareness and appreciation, but activism to protect beauty, ecological functions and other values. Mountain wanderer and Sierra Club founder John Muir is the progenitor, advocating for creation of national parks and in defense of forests. His poetic pleas to save Hetch Hetchy Valley and the redwoods remain moving, an expression of terranexus through efforts to protect places. This approach has included late twentieth and early twenty-first century authors as diverse as eco-anarchist Edward Abbey and climate change activist McKibben. Perhaps this second wave of conservation consciousness reached its zenith with Rachel Carson's *Silent Spring*, the 1962 blockbuster about the toxic dangers of pesticides that launched modern environmentalism and enabled us to clearly see the connection between human activity and the health of our landscape.

A third wave that introduced ethical conservation consciousness began with forester and ecologist Aldo Leopold, most well known for *A Sand County Almanac*. His land ethic "changes the role of Homo sapiens from conqueror of the land community to plain member and citizen of it," advancing terranexus via both a practical and moral connection of people to places. Integral to that view is an ecological way of thinking. "Land, then," he wrote, "is not merely soil; it is a fountain of energy flowing through a circuit of soils, plants and animals." Though once termed a "subversive science," an ecological worldview is today widely accepted though, unfortunately, less well practiced. Its best contemporary expression is in the writings of Kentuckian Wendell Berry who knows "Nature is party to all our deals and decisions, and she has more votes, a longer memory, and a sterner sense of justice than we do." The same sense of ecological connection flashes through the works of essayists Scott Russell Sanders and Annie Dillard. In fact, there is some element of Leopold's enviroethics in all the best writers that have come after him.

Leopold's land ethic imbued with terranexus could herald a fourth wave of conservation consciousness, encouraging us to prize not just pristine and magnificent places—the Yellowstones of the world—but more mundane precincts. A sense of terranexus encourages us to consciously explore rather than sleepwalk through areas where we live, work and visit. Exploration entices us to learn more, and as we come to know such places better we increasingly learn to appreciate them, hopefully even come to love them. And to the extent we appreciate and love them, we'll recognize our connection and want to invest our time, energy and money to protect and make them better. As we come to value familiar places, distant wonders will only grow in our esteem.

Garnering inspiration from mundane places to protect singular ones may sound far-fetched, but it is exactly in the Leopold mode. Though Leopold worked and lived in some of the American West's most fabulous country, it was a patch of worn-out land in an ordinary Wisconsin sand county on which he most lavished his love and found inspiration for world changing ideas. Leopold explained that "the [human] individual is a member of a community of interdependent parts." Likewise, terranexus erases, or at least blurs the lines between human and wild communities, between the built and natural environments, recognizing their linked ecology because "man is, in fact, only a member of a biotic team."

If ecological science teaches one basic lesson everyone readily understands, it's that all things are connected. But the connections we perceive at the scale of a wolf's territory or the habitat of bog turtles, a skunk cabbage filled wetland, and even at a watershed level, tend to be forgotten with regard to humanity, whether in rural areas, suburbs or large cities. Even among those who have elevated Leopold to near-sainthood, there seems some amnesia over the fact that he first explained natural communities by reference to their human analog, places where man's "instincts prompt him to compete for his place in that community, but his ethics prompt him also to co-operate."

Despite notable exceptions, the world seems to be divided into those who enjoy the pulse of urbanity and those who seek the flow of nature, those who often find the built environment too frenetic and ugly, and those for whom the natural world is dull and frightening. It is a chasm at least as wide as that perceived between science and the humanities in C.P. Snow's famous 1959 lecture, "The Two Cultures." Perhaps we need a deeper, more ecological way of looking at the world requiring slower, more deliberate movement through the landscape than we commonly experience. Sometimes we need simply to walk.

Many great conservationists have been inveterate walkers. After all, afoot is often the best means of traveling in wilder places. Inasmuch as what we perceive is often inversely proportional to how fast we move, going slowly enables us to see and understand more, essential conditions for terranexus. Furthermore, walking is a great stimulant to thought. "The rhythm of walking generates a kind of rhythm of thinking, and the passage through a landscape echoes or stimulates the passage through a series of thoughts," writes Rebecca Solnit in *Wanderlust*.

It comes as no surprise that Thoreau, Muir, and Teale are among many famous walkers who have shaped our understanding and love of nature. But the same is true of many great thinkers and lovers of urban areas such as twentieth century social critic Lewis Mumford who won the 1962 National Book Award for *The City in History*, or Alfred Kazin, a literary critic of the same era whose *A Walker in the City* fuses personal experience with acute observation of architecture and the human ecology of neighborhoods. Among contemporary writers, Harvard's John Stilgoe continues to explore city and suburb by foot and bicycle.

Terranexus teaches that built places can be just as beautiful, intriguing and curious as natural marvels. And a mix of the two often provides the most fascinating environments. Perhaps this is why my sense of terranexus first emerged in the White Mountains where both nature and human works stun the senses. Nature is awe inspiring, but the huts, well engineered trails, and features like the Mount Washington Cog Railway are critical to the experience.

City walkers are in the grand European tradition of the *flaneur*, the somewhat aimless wanderers who know urban areas with their raw senses, by footstep. Careful observers like naturalists, they distill stories from the architectural features of buildings whether it be a brick bond pattern, plate glass, or neon sign. They know how sunlight and shadow play on a plaza at various times of day and changes with the seasons. When a street is dug up they notice remnants of barrel stave water mains and cobblestone pavement. To a *flaneur* the past is always present. They know the stories of former inhabitants and watch the habits and moods of people in shops, on the street, and in cafes.

Nature and the world of human structures are inextricably linked. It's not civilization that is artificial or unnatural, but the distinction we have drawn between them. We need to extend hiking trails through cities, connecting them to wilder precincts. I'd like to see the blazes that take us up mountains and along bucolic waterbodies also mark paths through urban areas as does Boston's Freedom Trail. The New England Trail, newest of the eleven national scenic trails established by Congress does just that on its way from Long Island Sound toward New Hampshire's Mount Monadnock. Starting in Guilford, Connecticut, the first few miles take hikers past a veritable timeline of buildings representing almost 400 years of European settlement, including New England's oldest stone house.

The synergies of nature and culture first struck me while walking Boston's Emerald Necklace, a ten mile long series of connected green spaces that snake along the city's backbone from the historic Common to Franklin Park. A few years before paddling the Naugatuck, I discovered the route by accident when opening a crease-worn map to direct a friend to a Red Sox game. For the first time, I noticed swatches of deep green in the grid of city streets.

Much of the Necklace and the concept of making connections among city greenspaces was the brainchild of Frederick Law Olmsted, father of American landscape architecture, who is most famous for designing New York's Central Park. The route became my portal to discovering a rich mixture of history, art, architecture, birdlife, open waters and forest. But it wasn't all bucolic beauty. I also found pollution, invasive species, graffiti, a man urinating in the bushes, and broken walkways and bridges.

Each park along the necklace was a jewel commemorating a different aspect of Boston. With its walkways and tree shaded lawns the Common memorializes famous people and events with statuary and plaques. These objects commemorating the past were like tangible shadows in a place busy with tourists, concerts, and ballgames. From the Common, I entered the well planted flower beds and carefully groomed grass of the Public Garden which exuded a traditional Bostonian atmosphere of restrained formality. On the Commonwealth Avenue Mall, a wide linear swath of green dividing lanes of traffic, tree colonnades formed a leafy tunnel between low cliffs of brownstone and brick row houses lining the street.

Walking through the linear Backbay Fens and Riverway I passed bodies of water that were once tidal marshes and then fetid dumps for sewage. Today, Olmsted's green spaces are interrupted by roads, and parts of them are given over to athletic fields, community gardens and other uses. Nevertheless, I saw painted and snapping turtles basking in the sun, green and bull frogs hunched at the water's edge, and lots of birds from kingfishers to black crowned night herons, black ducks, mallards, and Canada geese. Children fished for carp, and held their catches high for me to see. Of course there was the inevitable half drowned shopping cart, some beer and soda bottles and other detritus in the water. Somewhere in this mixture of nature and trash the hopes and regrets of urban life seemed unwittingly expressed.

The Emerald Necklace continued to the open horizons and beautifully reflective waters of Jamaica Pond, a former reservoir, and then to the Arnold Arboretum, a place of both natural beauty and scientifically planted trees from every temperate region of the globe. I finally arrived at Franklin Park whose connection to the rest of the Necklace is tenuous because an Olmsted proposed parkway was never built. Large for a city park, its over 400 acres contained seemingly wild forests dense with maple, oak, and beech. But there were also invasive plants common in developed areas, like barberry and Japanese knotweed. There's a meadow disguised as a golf course, and a zoo. I watched red-tailed hawks wheeling overhead and wandered through places where not a single building was visible.

The Emerald Necklace is no wilderness, but perhaps it has sufficient wildness, as Thoreau would have it, to help preserve the world. Although nature purists might emphasize limitations, I felt exhilaration in finding nature and culture so well married. In Boston, an urban dweller does not have to flee the city to walk slightly on the wild side. If a traditional urban environment had these places, I wondered about the possibilities in less thickly settled areas. It seemed that preserving and promoting such nodes of wildness close to where people live could both enrich daily existence and save truly wild areas for special visits, thus avoiding their overuse. We need to be naturalists of urbanity, *flaneurs* of the wild.

In an age where preserved natural places like the White Mountains and Great Smoky Mountains National Park are often shrouded in air pollution and eaten alive by invasive species that come in the wake of human commerce, treating cities and wild places as distinct, unrelated islands seems particularly puzzling. How many people realize that the location of cities, housing tracts, agricultural fields, factories, and even the contour of roads are often largely dictated by bedrock and glacial geology and other natural factors? And recent books like David Owen's *Green Metropolis* and Edward Glaeser's *Triumph of the City* make a compelling case that if you care about nature, city living is preferred because urban dwellers require less living space, use less fuel, produce less trash and spend fewer hours in cars. Spreading people thinly across the countryside, Owen writes, "may make them feel greener, but . . . increases the damage, while also making the problems they cause harder to see and address."

Perhaps the link between built and natural environments was most elegiacally captured by journalist Howard Mansfield in his book *The Same Ax, Twice*. "The house is more like a natural landscape," he writes. "You are looking at time. Seven generations of life represented by a notch on a girt, a paint chip on a summer beam, the way the head of an adze met the wood one winter day in 1664. Life flowed through here and like a glacier left its marks upon wood and plaster." Such connections are the very soul of terranexus.

III

Wounded Places

For many of us, our appreciation of landscapes and an impulse toward terranexus is most intense where nature is dominant, from mountains to beaches, black spruce bogs to floodplain forests. They seduce us with their distinct beauty and refreshing differences from our usual haunts. But to be fully aware of and appreciate this world, our attachment must extend to working landscapes of farm and forest which provide food and fiber and offer a rich patchwork of trees, pasture, and cropland. It must embrace historic colonial towns with their broad greens, and nineteenth-century mill villages with their fortress-like factories. It needs to include city neighborhoods, commercial and industrial districts, both those that are well maintained and those suffering neglect. All these places help explain who we are as a society and as individuals. To see only those that please us is to inhabit a Disneyland of our own making.

While they need no encouragement or protection, even strip developments with their garish signs, confusing traffic lanes, fast food restaurants, filling stations and big box stores are functional parts of our landscape along with interstate highways and housing tracts. We may think some of these landscapes ugly and wasteful, but few of us do not use them. And I admit to feeling an occasional thrum of excitement in the busy commerce of a strip mall, or the cinematic movement of topography at sixty-five miles per hour. All these are areas in which we live, work and do business. They are places that deserve our interest and attention. The more we understand them, the more we will know what is needed to create better places.

Regardless of our attempts to minimize and see some value in less than ideal places, there will always be landscapes severely compromised in support of a civilization that also builds meaningful cultural spaces and protects the beauty of nature. Places sacrificed for the less glorious needs of civilization include those chosen for waste disposal, energy development, mineral extraction, and other uses that severely damage the land. I'm reminded of F. Scott Fitzgerald's "valley of ashes" in *The Great Gatsby*, a "desolate area" of "gray land" with "spasms of bleak dust which drift endlessly over it."

After shrinking the need for them as much as feasible, do we justify such places as necessary evils supporting a greater good? Do we try seeing a stark beauty in their barren and often toxic precincts? Sometimes poisoned places, of which Chernobyl is the supreme example, become accidental wildernesses, opportunities for a revival of nature, though at great human cost and unknown long term impacts to the plant and animal communities that recolonize them.

Ultimately, we must acknowledge our culpability in devastating some parts of the planet. Perhaps the best we can do is maintain consciousness of their pernicious impacts and hope such awareness motivates us to limit these zones. "Certainly these sad, toxic, taboo places deserve as much recognition and gratitude as their unmolested counterparts," Trebe Johnson argued in a *2015* issue of *Orion* magazine, about a slice of bucolic Pennsylvania sacrificed for hydrofracking. "By offering a bit of beauty," she writes, "to a being or place that has been felled, fracked, polluted, abused, or in some other way robbed of its dignity and purpose, I can replenish its loveliness. By believing—and then acting on—the conviction that a place is worthy of receiving some kind of gift, my consciousness shifts from anger, disgust, or sadness to one of compassion, engagement and creativity."

Johnson's notion may seem somewhat romantic, but sometimes there's nothing else we can do. Although such places are not usually sources of creative inspiration, perhaps some of that loveliness can be replenished with art. Robert Smithson, most famous for monumental land art like his 1,500-foot-long Spiral Jetty extending into Utah's Great Salt Lake, chastised artists who retreated to scenic beauty spots rather than try to form a dialectic between people and nature. As early as the 1960s he saw the industrial landscapes of northern New Jersey as worthy of artistic expression.

Maybe the best way to express gratitude for the places we have wounded is to repair those that can be healed. This may be a cleaned brownfield put to new economic use or the site of a demolished building that can be reborn as a garden or pocket park. Even landfills can be transformed.

The old Hartford landfill, a mound that rises about 130 feet above the surrounding landscape, is not far from my home and among my favorite degraded places to visit. Sandwiched between Interstate 91 and the Connecticut River in the north end of the city, it accepted garbage and incinerator ash from 1940 to 1988. Now the grassy mound is healing. Wells drilled into the interior have produced methane gas to generate electricity and a solar panel array has been installed. In season, the uneven plateau sports a startlingly lustrous prairie of waist high grass where daisies, lacy yarrow, buttercup, white and red clover, hawkweed, the purplish pea-like blossoms of cow vetch, and many other common roadside flowers bloom profusely. The grasses and sky resound with avian life. When I was last there, redwing blackbirds cruised just above the meadow and uncommon savannah sparrows alighted on fencepost-sized gas wellhead pipes. A spotted sand-piper with a clutch of young crossed the gravel wheel track, and a killdeer scurried away performing a fake broken wing routine. A great blue heron flew overhead with its long legs outstretched and red-tailed hawks circled far above. There was a kingbird, indigo buntings, orchard and Baltimore orioles, grackles, cedar waxwings whose tail-tips looked as if dipped in yellow paint, goldfinches and other birds. It was a giddy carnival of color and song.

A tour bus operator recently expressed interest in excursions to the landfill's summit for what may be the finest panorama along the banks of New England's longest river. To the north and south, the grand corridor of the Connecticut valley spreads out like a map with trees, fields, water, roadways, church spires, smokestacks, bridges, rooftops, and all the tackle and infrastructure of modern life visible for miles. To the west and east the horizon rises to lumpy, rugged highlands punctuated by a few antennae and towers.

Once a smelly eyesore, this accidental high meadow is becoming a place of intriguing beauty, source of energy, and reserve of biodiversity. Bird expert Jay Kaplan calls it the best grassland habitat in the region. Here is recycling writ large, a debauched and degraded landscape transformed and healing into something else, a spot with a second chance.

It's not possible to restore the swale where Hartford's old landfill now rises, but we might not want to do so even if we could. New opportunities abound for this and other mounds of used up and broken stuff. Less than an hour's drive away, the old Milford landfill—rising fifty feet above the beach at Silver Sands State Park—offers spectacular views of Charles Island and Long Island Sound. When finally developed, visitors will be able to hike or drive cars via a narrow roadway that will spiral around the landform to a summit featuring a pavilion for picnics and weddings. Old landfills have been recast as athletic fields, parking lots and even commercial building spaces. These places may be ugly on the inside, but can be made to look beautiful and become meaningful. We cannot afford to forget their existence or our connection to them.

Landfills are only one among many wounded places that we should remember. While rivers like the Naugatuck have been reborn after dying the death of a thousand polluting discharges, others have been buried because they flooded or pollution made them disgusting. Sometimes they used space that myopic developers and politicians found more suitable for roads and buildings. These rivers may be gone, but not always forgotten. Providence, Rhode Island exposed and moved about two-thirds of a mile of river in the heart of downtown in the 1990s, watercourses that had been sealed for decades beneath roadways and parking areas. The rivers have become an economic driver and tourists flock to walk the tree lined cobblestone paths alongside them, cross the graceful Venetian bridges, take boat rides, and see the world famous WaterFire display with its 86 burning braziers. Experiencing terranexus, caring people connected with Providence's waterways and helped create the city's amazing revival.

Terranexus demands we declare war on the kind of cultural dementia that causes us to ignore such places. True lovers of landscape see not only today's prospect, but imagine what once was and what might yet be. They make the places where we dwell more livable. We must explore forgotten and forbidding places so we can tell their stories. Such places beckon with unexpected adventure.

Like Providence, Hartford also has a buried watercourse. The Park River is invisible in the last two miles before its confluence with the Connecticut

River. Its waters are never pelted by raindrops, ruffled by wind or dappled with sunlight. Most people are unaware a river runs through the city because it runs beneath it. The Park was buried by the Army Corps of Engineers between the 1930s and 1980s, punished by entombment in massive concrete conduits for the sin of destructive floods caused by poor development in its watershed.

For years I passed the twenty-seven-foot-high tunnel opening on my way to work. At last curiosity about its course got the best of me. Although in high water the trip could be extremely dangerous, I found that low water offered a peculiar paddle through an ersatz wilderness devoid of people, roads, buildings, sun, and sky. It was an eerie, dank world of echoes and eternal night that enfeebled even the brightest flashlight.

Colorful graffiti marked the entrance, but light quickly faded until reverberating sounds and a swampy smell dominated the senses. Riffles around occasional rocks and a little water seeping from above were as loud as waterfalls. Glancing around with a headlamp I saw smooth masonry walls punctuated with metal capped sluice gates and screened intakes. Although the darkness was daunting, the trip was remarkably uneventful. No filth floated on the water; there were no clots of garbage or unexpected rapids. Only a small school of catfish clustering around the boat near the tunnel's end was noteworthy. Ultimately, I emerged from the disquiet and constriction of darkness into an eye-squinting explosion of light and space on the Connecticut River, the city's skyline behind me.

This river may never become a favorite paddle of canoeists. But the underground experience holds a bizarre allure that mixes our fascination with oversized engineering feats, fear of darkness, attraction to moving water and desire for offbeat experiences. A local outfitter was once interested in leading commercial trips, but city bureaucrats and risk managers axed that idea. Thus, the Park River remains an outlaw journey, perhaps amping its allure. It may be out of sight, but should never be out of mind. This forever dark reach of river has much to teach about the ultimate price we pay for abusing our waterways, about engineering, development, public works, and history.

IV
Deep Traveling

How can we experience teranexus and narrow the divide between wild and built landscapes? How will teranexus revive interest in the conservation of unique natural and cultural places? Only by deep travel, a form of practiced, concentrated observation amplifying what we see close to home.

The commonplace world around us brims with stories that deep travelers detect, whether they are found in the distant geological record of bedrock displayed at roadside rock cuts, at cemeteries, or in street names. Our landscape is dotted with revealing clues that enable us to read what has happened and what might be. Deep travelers know this and see the world in high definition.

Deep travelers go beyond what is visible and unlock the stories hidden in people, places and phenomena. They see connections and narratives everywhere. Deep travelers connect the dots among diverse phenomena such as swift streams, nineteenth century waterpower factories with worker housing, and twenty-first century neighborhood ethnicities. A deep traveler perceives changes over time, observes history in roadside milestones, building materials, abandoned structures, and the tales people tell on the street, in coffee shops and barrooms. It's not merely a matter of improved visual acuity, but transfigured vision that injects awe and wonder into everyday experience. Deep traveling leads to intimate place knowledge. Knowledge begets understanding, and understanding leads to terranexus, profound connection to places both singular and ordinary.

Traveling deeply requires no mystical sixth sense or specialized academic training, only sensitivity to the simple language of the landscape. The words are the details we see every day, from big trees to swayback barns. The grammar is found in the way we weave those isolated details into sentences and paragraphs of meaning that create narratives. No artifact, whether a stone wall or a housing tract, is an isolated word or sentence in the story of people and landscape.

People who care about places must be deep travelers. We are obligated to tell stories that will excite those blind and deaf to the everyday magic animating both ordinary and unique areas. No place, not a polluted wasteland nor a seemingly boring subdivision is without compelling tales and fascinating human and landscape confluences. Those who cannot see these links need to look harder or differently. We must be cartographers of imagination, drawing what unconventional travel writer William Least Heat-Moon calls a "deep map," an amalgam of topography, history, biography, folklore, politics, geology and natural history.

Deep travel heightens awareness, creates a landscape consciousness that results in terranexus. Such purposeful looking is more than a windshield tour or a plat-like view from a plane. It is often unplanned, serendipitous exploration. "Ordinary exploration begins in casual indirection," writes John Stilgoe in *Outside Lies Magic*, "in the juiciest sort of indecision, in deliberate, then routine fits of absence of mind.

Walk three quarters of the way around the block, then strike out on a vector, a more or less straight line toward nothing in particular, follow the downgrade or the newer pavement, head for the shadow of trees ahead, strike off toward the sound of the belfry clock, follow the scent of the bakery back door, drift downhill toward the river."

Opportunities for deep travel exploring exist all around us and only await our imaginations to match the rich complexity of the landscape. Few realize, for instance, that cemeteries are as much for the living as for the dead. Beyond being places to pay respects to our predecessors, graveyards are schoolrooms of history, sculpture, genealogy and natural history. They play host to endangered species, and are often ideal places to observe birds, wildflowers, mushrooms, butterflies and other natural delights.

In old industrial precincts we sometimes find power and transportation canals that take us to the very intersection of nature and invention. They can be repurposed for boating, fishing and the sheer enjoyment of walking beside their waters. Lowell National Historical Park in Massachusetts, a cradle of American textile production, has long made use of these artificial rivers for the education and sheer enjoyment of the public.

Abandoned railroad rights-of-way repurposed for non-motorized travel, like the one that leads to my listening point, increasingly take us to experience nature, even in the heart of urban areas. The elevated railway known as the High Line on Manhattan's west side, built for industrial freight, has been reengineered into a walkway above city streets. Now filled with naturalistic plantings and birdsong, it offers spacious views of the sky and horizon, pleasures that are sometimes at a premium amidst city towers. Striding along the route is energizing. Simultaneously I feel the galvanizing movement of the city and the soothing elements of nature.

Deep travelers know that ruins can add drama, beauty, and meaning to our landscape. Tumbledown structures like mills, grand houses, and public buildings are not necessarily eyesores or signs of blight. Properly curated, they are opportunities to enrich our culture and provide economic development. They can instill community pride and prepare us for future change by illustrating time's passage.

Nineteenth century painters like Thomas Cole and writers like Washington Irving lamented that this nation lacked ruins to chronicle the past and give depth to its countryside. Though the most famous of Hudson River School painters is best remembered for canvasses depicting the grandeur of America's natural scenery, Frederic Church traveled to Greece, the middle east and elsewhere to capture the moldering remains of old buildings. Today this deep traveling genius of American art could have stayed home and painted abandoned factories, some of which stand like noble castles along our waterways.

Not every crumbling mill and other structures ought to be saved as ruins, but it's time we realize that restoration and demolition aren't the sole alternatives. Those who think stabilized ruins don't add pecuniary value should talk to British tourism officials about Roman remnants and the remains of abbeys and towers. Closer to home, the Eastern States Penitentiary in Philadelphia is a preserved ruin attracting thousands annually, as are the brick remnants of a Barboursville, Virginia house designed by Thomas Jefferson, and the cliff dwellings of Colorado's Mesa Verde National Park.

Architectural ruins demonstrate nature's relentlessness as a solvent of all humankind creates. They evidence our constant battle against the forces of entropy. Not every example of picturesque decay is worthy, but we will be remiss if we fail to make an effort to protect and interpret those that can enliven our landscape by visibly and tangibly telling a story of ingenuity, hard work, and other transcendent values.

Renewed enthusiasm for natural and cultural resource conservation depends on deep travelers, people of insight who care about places near and far, to light fires of enthusiasm under those around them. We can talk about the magnificence of distant regions, but we will be most effective if we start generating more interest in familiar spots, like Leopold did for his beloved patch of tired sand county. The first step is to steep ourselves in the natural and cultural history, science, legends, and lore of the everyday areas where we live, work and play. We must tell stories that give meaning to those places. This is the essence of deep travel, the heart of terranexus.

The need for balanced, diverse, healthy functioning systems transcends differences between built and natural worlds. Most times we experience synergies between the two. "What we long for," wrote the eminent biologist Rene Dubois, "is rarely nature in the raw, more often it is a landscape suited to human limitations and shared by the efforts and aspirations that have created civilized life." Paradoxically, revival of the conservation movement and protection of our most magnificent places lies in understanding and appreciating those areas where natural values are not necessarily pure or ascendant. It's not just a matter of demonstrating that the economy and culture of cities are codependent on the complexities of natural habitats, but of valuing a relationship regardless of similarities and distinctions. A sane, sustainable future demands ecological thinking in the built environment and understanding of how culture affects the natural world. It demands better connection to the everyday landscapes around us. It demands terranexus.

When in the woods we need to extol old mill dams, faded roads, farm walls, abandoned quarries, changing plant assemblages, and rock outcrops that illuminate the "why" of a place. Here in New England, for example, we must relate the story of primeval forest, Native habitation, colonial farming, early industry, reforestation, and development. While walking in our neighborhoods, we must be attuned to architecture that speaks of fashion, money, and technology as we recall quirky residents and past land uses. Street names alone often read like a completed crossword puzzle inviting us to pose the questions that they answer. We must make places personal by conveying that such stories are also our story, that we exist in the very continuum we describe. Richer places mean richer lives. The places where we live can be places that we love if we dare to make the investment—not of money, but of time and energy.

The deepest travel begins at home. Splendor awaits just beyond the doorstep.

About the Author

DAVID K. LEFF is an essayist, pushcart prize nominated poet and former deputy commissioner of the Connecticut Department of Environmental Protection. He is the author of five nonfiction books, three volumes of poetry and a novel in verse. In 2016-2017 the National Park Service appointed him poet-in-residence for the New England National Scenic Trail (NET). David's journals, correspondence, and other papers are archived at the University of Massachusetts Libraries in Amherst. He is the town historian and town meeting moderator of Canton, Connecticut where he also served 26 years as a volunteer firefighter.

WWW.DAVIDKLEFF.COM

LITTLE
BOUND BOOKS

SMALL BOOKS, BIG IMPACT

The Little Bound Books Essay Series
Personal. Poignant. Powerful.

WWW.LITTLEBOUNDBOOKS.COM

HOMEBOUND PUBLICATIONS

Ensuring that the mainstream isn't the only stream.

At Homebound Publications, we publish books written by independent voices for independent minds. Our books focus on a return to simplicity and balance, connection to the earth and each other, and the search for meaning and authenticity. Founded in 2011, Homebound Publications is one of the rising independent publishers in the country. Collectively through our imprints, we publish between fifteen to twenty offerings each year. Our authors have received dozens of awards, including: *Foreword Reviews'* Book of the Year, Nautilus Book Award, Benjamin Franklin Book Awards, and Saltire Literary Awards. Highly-respected among bookstores, readers and authors alike, Homebound Publications has a proven devotion to quality, originality and integrity.

We are a small press with big ideas. As an independent publisher we strive to ensure that the mainstream is not the only stream. It is our intention at Homebound Publications to preserve contemplative storytelling. We publish full-length introspective works of creative non-fiction as well as essay collections, travel writing, poetry, and novels. In all our titles, our intention is to introduce new perspectives that will directly aid humankind in the trials we face at present as a global village.

WWW.HOMEBOUNDPUBLICATIONS.COM

CPSIA information can be obtained
at www.ICGtesting.com
Printed in the USA
LVHW04s0513260518
578593LV00002B/9/P